YOUR KNOWLEDGE HAS VALUE

- We will publish your bachelor's and master's thesis, essays and papers

- Your own eBook and book - sold worldwide in all relevant shops

- Earn money with each sale

Upload your text at www.GRIN.com
and publish for free

Sonam Tobgay

Report on Visit to National Institute of Traditional Medicine and National Biodiversity Center. Systematic Botany

GRIN Publishing

Bibliographic information published by the German National Library:

The German National Library lists this publication in the National Bibliography;
detailed bibliographic data are available on the Internet at http://dnb.dnb.de .

Imprint:

Copyright © 2015 GRIN Verlag GmbH
Print and binding: Books on Demand GmbH, Norderstedt Germany
ISBN: 978-3-656-93768-5

This book at GRIN:

http://www.grin.com/en/e-book/295574/report-on-visit-to-national-institute-of-
traditional-medicine-and-national

GRIN - Your knowledge has value

Since its foundation in 1998, GRIN has specialized in publishing academic texts by students, college teachers and other academics as e-book and printed book. The website www.grin.com is an ideal platform for presenting term papers, final papers, scientific essays, dissertations and specialist books.

Visit us on the internet:

http://www.grin.com/

http://www.facebook.com/grincom

http://www.twitter.com/grin_com

ROYAL UNIVERSITY OF BHUTAN

COLLEGE OF NATURAL RESOURCES

LOBESA: PUNAKHA

A Report on Visit to National Institute of Traditional Medicine
and National Biodiversity Center, Thimphu

(SYSTEMATIC BOTANY)

Sonam Tobgay

B.Sc. FORESTRY, 2015

Table of Contents

Visit at National Institute of Traditional Medicine

The National Institute of Traditional is located at Kawang Jangsa, Thimphu. It was established in 1971 and upgraded in 1992. Today, the institute functions as a member Faculty of Traditional Medicine under the Khasar Gyalpo University of Medical Sciences of Bhutan. Currently it has Three Units: (1) The Hospital Unit/ Indigenous Hospital, (2) Faculty of traditional Medicine or the training Unit and (3) The Pharmaceutical and Research Unit.

Bhutan is known as the land of medicinal plants. Bhutan is regarded as one of the ten global hot spots in terms of biodiversity and environmental conservation. Apart from its rich biodiversity and natural resources, there are some endanger flora and fauna. The Bhutan Landscape habitats more than 600 medicinal plants identified, and at least 300 of these are commonly used by local practitioners in the country for preparing local medicines.

Productions in the past, all traditional medicines were produced manually. Small scale mechanized production started only in 1982 with support from World Health Organization but local healer has been practicing from centuries and yet still exists. Although most of the plant ingredients are collected from wild, some of the species are now being cultivated/ domesticated by the farmers. Perhaps medicinal plants collected are one of the first economical activities initiated and sustained in this highland of raw materials procured within the country. Some part of the remaining raw materials which is not available in the country is mostly imported from India.

Storage and distribution

The Traditional Medicines produced are being delivered in the Traditional Medicines center at different places of the country. The store management at the district level is fully integrated with the modern medicines. The procurement of equipment and other supplies are also done.

Herbal Collection Site:

Some species are introduced for cultivation, collection, Storage in collaboration with the Medicinal and Aromatic Plants. The focus is on the development of medicinal plants in all levels including sustainable collection, production and marketing of herbal products to function the Unit as a self-sustaining commercial entity. The collection medicinal plant sites are mostly from high altitude and low altitude.

Utilization of Traditional Medicines

The number of patients using traditional medicines is steadily increasing over the years. The system is quite popular among the elder population and the patients in the district hospitals are using traditional medicine.

Processing/Production of Tablets and Capsules

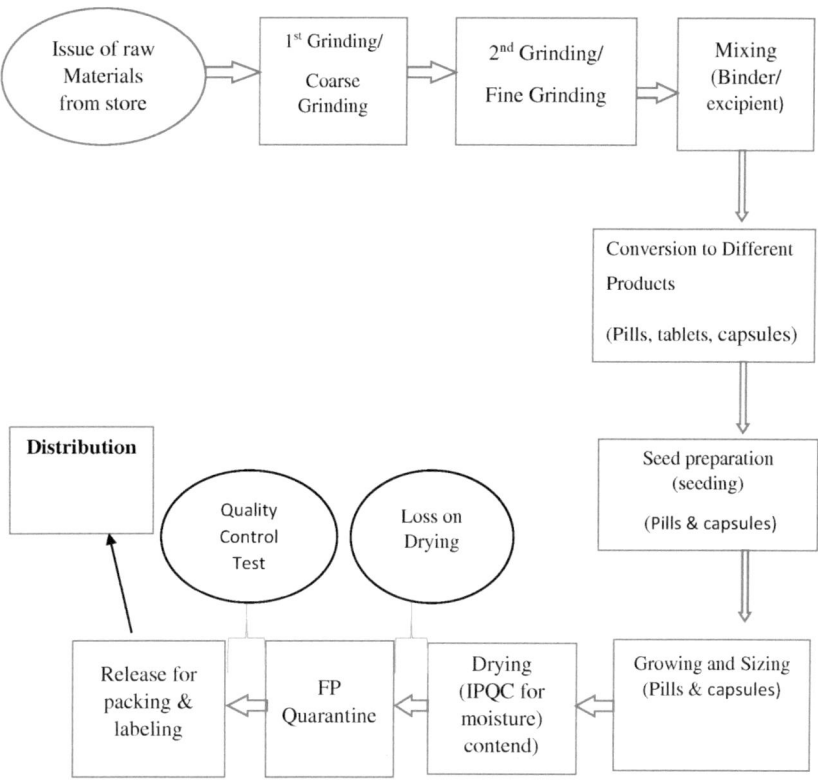

The Menjong Sorig Pharmaceuticals, otherwise known as Pharmaceutical and Research Unit is the premier manufacturer of the traditional herbal-based medicines supplied to all the Hospitals in Bhutan. A number of medicines are produced here in its large laboratories. The components of the medicines produced here are minerals, animal parts, precious metals and gems and plants. Products such as Pills, tablets, capsules, powder, ointments, syrups crude extracts and herbal mixtures are made.

The medicinal plants that are used in National Institute of Traditional Medicine are briefly described with botanical names, family, part used and Altitude range of individual medicinal plants. A dendrogram for these 10 medicinal Plants is also constructed using APG III system of Plant Classification.

SL. No	Botanical name	Family name	Altitude	Parts Used	Used For
1	Aconitum orochryseum	Ranunculacea e	3900-4800	Aerial parts	Anti-dote for snake bite, Allays common cold, bile, inflammation and dysentery
2	Phyllanthus emblica	Phyllanthacea e	100-1400	Fruits	Anti-inflammatory, antiulcer, hepatoprotective, and anticancer actions.
3	Carum carvi	Apiaceae	2500-4150	Seeds/fruits	Allays giddiness, Hypertension, poisonings and eye infection.
4	Berberis aristata	Berberidaceae	1670-3000	Inner bark	Anti-dote in case of poisoning, chronic cough and cold
5	Terminalia chebula	Combretaceae	100-1400	Fruits	Cough and cold,
6	Inula grandiflora	Asteraceae	2000-3300	Flowers	Cures abscess/boils, numbness, fever and evil affliction
7	Terminalia bellirica	Combretaceae	100-1400	Fruits	Cough, relieves blocked phlegm, controls bleeding in the sputum. lowering cholesterol and blood pressure
8	Gentiana robusta	Gentianaceae	3500-4600	Flowers	Heals wounds, swelling and inflammation of stomach and liver

9	*Geranium procurrens*	Geraniaceae	3500-4200	Roots	Anti-diarrhoeal and anti-toxin. Allays cough and cold, bronchitis, swelling of limbs
10	*Rheum nobile*	Polygonaceae	4000-4600	Leaves and Flowers	Laxative, diuretic, anti-emetic. Useful for swelling and sensation of fullness in the abdominal area and helps retain body fluid

Dendrogram 0f 10 medicinal plants using APG system of classification

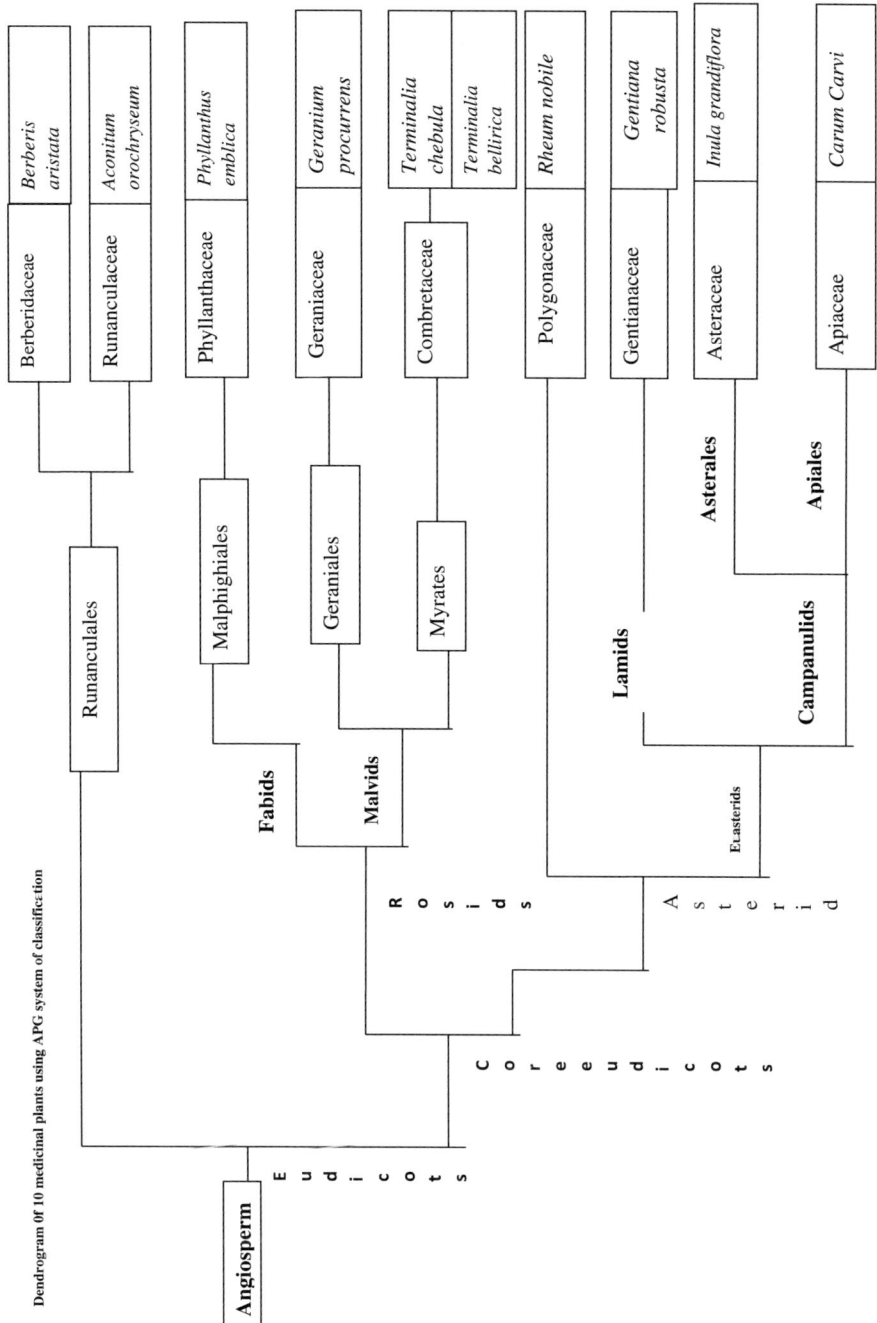

Visit at National Biodiversity Centre, Serbithang.

National Biodiversity Centre is also collaborates with National Institute of Medicines especially for identification of medicinal plants as well as publication of Journals and books. The National Herbarium is one of main section of NBC for Flora of Bhutan. The foundation of the National Herbarium dates back to the 1970s. In 1999, with the establishing of the National Biodiversity Centre, the herbarium was move to the present location at Serbithang from Taba. The new herbarium building at Serbithang was completed in 2002. The National Biodiversity Center has following Units: Royal Botanical Garden, Bio-prospecting, Plant Genetic Resources and Animal Genetic Resources.

The National Herbarium serves as the centre of botanical survey and exploration. Besides functioning as the repository of reference material, it also provides a means of identification of plants by matching unnamed plants with named specimens in the collection. The herbarium collections include Angiosperms, Gymnosperms, Pteridophytes and Bryophytes.

Current Functions and objectives

- Vascular plants collected, preserved and documented
- To Study the endemic plant species of Bhutan and possible impacts of climate change.
- Sample of artifacts
- To establish regional herbarium.
- To develop baseline data on ethno-botanical plants.
- To carry out gap analysis of herbarium specimens and published Flora.

Herbarium

Herbarium is a store house of dried plant specimens collected from far and wide, mounted on appropriate sheets, arranged according to some known system of classification and kept in pigeon holes of steel or wooden cupboards. With the successful completion of the "Support to the National Herbarium and Flora of Bhutan" Project, the National Herbarium has developed into a national reference centre for Bhutanese flora and related works. The herbarium specimens are now stored in steel cabinets. Electric ovens are used to dry specimens during wet seasons. A deep freezer is in place to treat specimens for ensuring pest and disease control. The herbarium has adequate working space for visitors and researchers. The preparation of Herbarium involves following processes:

1. **Collections** –Specimens should be healthy, with at least some fully expanded leaves with flowers, fruits and vegetative parts and in smaller plants including underground portions (roots/rhizomes/ bulb). Press immediately in the fields or collecting in polythene bags is another option. Maintain field notes to record information about plant, habitat, locality and associate plants, etc.

2. **Pressing specimens** – Press it right away in the field using newspapers and wooden press to produce best specimens and also this prevents fungal infections and preserves Color. Then dry the press with specimens in the sun and change newspapers daily. Use blotting papers in between the specimens to absorb water from thick and succulent plants

3. **Drying** – dry under the bright sun or in microwave oven maintaining the temperature below 55°C. Use cardboard papers for proper aeration

4. **Identifying and mounting specimens** - use guide books, illustrations, pictures, Flora books – mount specimens on sheets after accurate identification using electric device, glue, thread and needles.

5. **Labelling** – prepare label according to the field notes. It should have botanical name, locality, habitat, description or habit, altitude, collector's name, collection number, date etc

6. **Cataloguing** – Finally, place Herbarium sheet in the Genus cover and catalogue specimens in cabinets according to Engler and Prantle system of classification.

Conclusion

Visit to National Institute of Traditional Medicine (NITM) and National Biodiversity Center (NBC) was very fruitful in term of learning the medicine usage, processing and the availability of plants in the country, bio-prospecting, herbarium techniques and new system of classification of plants. The major challenges for traditional medicine services are to mobilize adequate resources and all stakeholders should also be aware of endanger species and other plants sustainability. The National Biodiversity Centre is house for many programs like Plant Genetic Resources, Animal Genetic Resources, Bio-prospecting Program, Biodiversity Information System, Royal Botanical Garden and National Herbarium. The NBC is giving importance on preserving the Native Agri-Crops Seed Bank, Native domestic animals semen, rescuing wild plants and Plant diversity collecting specimens. There is still room for the improvement of existing programs as well as upcoming Programs with Project.

Reference

Cole, T. C. H., & Hilger, H. H. (2011). *Angiosperm Phylogeny Flowering Plant Systematics*: tolweb.org/ (http://www2.biologie.fu-berlin.de/sysbot/poster/poster1.pdf).

Fculty of Traditional Medicine,Khaser Gyalpo University of Medical Science of Bhutan Retrieved from http://www.nitm.edu.bt/

National Biodiversity Center, Ministry of Agriculture and Forest Retrieved from
http://www.nbc.gov.bt

Wangchuck P, Samten & Ugyen. (2009) *High Altitude Medicinal Plants of Bhutan: An illustrated Guide for Practical Use*. Pharmaceutical and Research Unit, Institute of Traditional Medicine Services, Ministry of Health, Thimphu: Bhutan